Cursed Objects

Cursed Objects
Jason Christie

Coach House Books | Toronto

Published with the generous assistance of the Canada Council for the Arts
and the Ontario Arts Council. Coach House Books also gratefully acknowl-
edges the support of the Government of Canada through the Canada Book
Fund and the Government of Ontario through the Ontario Book Publishing
Tax Credit.

LIBRARY AND ARCHIVES CANADA CATALOGUING IN PUBLICATION

Title: Cursed objects / Jason Christie.
Names: Christie, Jason, author.
Description: Poems.
Identifiers: Canadiana (print) 20190050454 | Canadiana (ebook) 20190050462
| ISBN 9781552453841 (softcover) | ISBN 9781770565883 (PDF) | ISBN
9781770565876 (EPUB)
Classification: LCC PS8555.H719 C87 2019 | DDC C811/.6—dc23

CONTENTS

'I offer the public pure cold water.'

– Jean Sibelius

'An object is just a slow event.'

– Stanley Eveling

'From the far end of an infinite cold black tunnel a helden tenor rings like a quasar – "du bist meine Augen..."'

– Maxine Gadd,
'Loon,' Subway Under Byzantium: Poems, 1988–1996

NOISE

plu

   m

            met

       plu

  m   .

      met

         plu

  met lu

pl ummet

   plu

                      p

    lu

      u

m

        m

    et

      i   p

      l

 u

    plum e

   mm

                   et

  etpluet   s

       plu

u

  etetet

           err

   plu   mett

  s

lèse-majesté

a self-guided tour of follies and environs with a crowd moving as one

PART 1

A VIRTUAL WALKING TOUR OF VARIOUS FAKE ENGLISH-LANGUAGE BUILDINGS

'who said the romance in
follies has gone'

> – Andrew Plumridge,
> The Folly Fellowship Newsletter No. 1

'You must put your ruin at last into the hands of nature.'
> – William Gilpin, *Observations, Relative Chiefly To Picturesque Beauty, Made in the Year 1772, On several Parts of England, Particularly The Mountains, And Lakes Of Cumberland, and Westmoreland*

BROADWAY TOWER
Worchestershire, England

| |

    , A
fall without reference for
words unabridged rain

| |

| |

| |

| |

| |

| |

referents adrift at night
nouns like, rightmost path,
warm West Wind, warmth
dusks from within a book
to comfort me to sleep –

| |

| |

| |

    what do I do with my eventual
self, my buildings: protect me!

| |

| |

and if I'm asleep or if I am to sleep
then what arrogant history rises as
if to say, smile aslant, hey you how's
that whole four-walls-and-a-roof
thing working out for you? I cling to

| |

| |

platitudes and now, as if to say
I'm hiding inside my body,
a device rises, or the poem
appears on a blue screen; a
piece of the poem lodges itself
in my mind and I find the words:
a rectangle around me or my program,
fade out to a blue sky with clouds
bounded by chrome, a green hill rolls
and music softly chimes as if to
say welcome home, windows,
some doors open, earth or wood
floors, a circular vent, cut in
rock, a rock ceiling, a stone cover,
a knight rolled away a waylaid
jamb – at dawn it began, undone
energy escaping my house as
sound, a spark, tilting words
reading – bring wind and rain
aslant like sharp notes to the C
blasting my windows, gales
threw landmarks askew
heroic aghast, away, listing
somewhere on the internet
windmills to forgetfulness

I survive weather warnings
even as crude symbols accrue
weight, I dissemble, repeating:
crumble and build, crumble
and build
              another system
of signs around my body
because the old one retired

| |
| |
| |
| |
| |
| |
| |
| |
| |
| |
| |
| |

electric arc flung high,
skyward, a dark, nuclear
dry, electric arc, singe
songs slow down to dawn
but until then, I'm under-
gone, gaunt, an engine
guttering low, slowing
into a cycle of haunt
and reclaim, my life is
a dream at the end of
archiving pavement,
solid yellow line, solid

white line, dotted line,
some unclaimed taunt rests,
neat metal rows, like a petal
peeled to the metal, in a lane
of white beams, lonely as a clown
lowering down the escarpment like a road,
cars gleam the curb, spawn free shapes,
dark amber around me, breath drawn deep,
shadows enumerate a floating street number
glowing in a font I don't recognize, a sign saying:
at dawn water draws into a stem, a stem saying

       a full street glows for me
       the moon curves away, C

        | |

       moon filling or fallow, lead shining
edge, landslide rumbles past, then clouds
settle into more land spreading below
the dark, my device contemplates me as
       I rise to find earth, tides
       in their depth and height,
       trough and peak, lengthening,
       valley and crest, I rise afield
       and in rising, I find light
       becoming a landscape
       which is not the opposite
       of the darkness I was expecting

        | |
        | |
        | |
        | |
        | |
        | |

wood smoke, unnatural cloud, plummet and
abundance, contrary to a forest's fiery
decline, a line stretched afield all combusted,
taut, and compromised, all blended
wires, all light transit or movement
to scorch, rally for collection as a forest of
pixels on my computer monitor, dark
edge across bright expanse to dark fiery edge
or fire cuts into dark at night or by day
it's all the same to me, on the other side
of the glass, online with great effort, curbing
all phasic slumps or hedging growls, a notch
where all else fails against data, and access fails
against firewalls all set around my devices, my
location, friends and strangers, attend to me
pinging for everyone on a solo map I created:
the password is the answer to the security
question I set when I was seven – *what is
your favourite animal?* – the camera clicks –

> a line stretched, technical frame
> taut, prized articulations –
> plunging ice line or gala regalia –
> an economy of rising tides failing
> – the ridgeline plummets, we fall
> – surfacing the economy with
> bright, new paint – such trends
> stall at becoming, and gutter or become
> pointing, then more public pointing

weather occurs for
narrative embellishment:
it storms, it rains, it is cold

where people gather together and
repeat cycles of smoke and fire,
       stocks of smoke and fire plummet and
an economy of wood and spark rises,
an abundance of ash profits the fire tender
       as I shake my body into another's
       unaffordable river view, or
       desirable narrative, each forest
       forward thinking, and alphabetic.

              | |

       I'm alone at the end, far from whatever
       primordial energy built a home and
       then escaped it, whatever body
       found itself and then forgot to light
       beacon fires upon the hill for all to see

              | |

              | |

              | |

              | |

              | |

              | |

              | |

              | |

              | |

              | |

              | |

THE CASINO AT MARINO
Dublin, Ireland

          bells destroyed
                    mid-ring, the artist
                              ,wrought
iron bells destroyed   , a tower
                              sounding
                                   town
surrounding
                    hills   ,   noise breaks
          intimate metal
                    flakes, a ring of
          rust ,
          red mesh rattle
                    patched toward
other colonies, the storm stretches

                         ,

                    as it happens :
          a slow
          hammering
          juggernaut

                                   , a

mess of arrangements between page
sound and score, poetry of the unseen
or inkblots following gentle, elaborate
                    curves, an
architecture left momentarily aloft or
alone dents the surface of the page, a
city, unwatched and fragile,
            wavers with prepositions:
letters like distracted sentries link arm in arm,
                    punctuation drifts into a line,
            sentences alight on screen like pillars asleep

                    retroactive wayfinding
                    weight-fraught bulk
                    waiting until dusk

                    at night I pretend to dream
                    whatever giant invisible dream
                    my generation is supposed to dream
                    while sleeping alone under the stars
moonlight creates walls for our
little house, our glorified room, I'm
slowing against my tendency
to heap trash on top of the pile
and catch it as it tumbles, slowing
against throwing it all back on top
and again the clock in our quiet room
makes bright blue lights of our life
a silent register, accumulating dead skin

                    to delay arriving at the same place
                    between lines of the poem, lines

on the forehead, a line and mirror
lines I've cast in bronze and struck,
underline these words: *a bell chimes*

an effect or sleight of hand not every mutated word appears to
reach through mottled tree trunks and mossy rocks to springs,
reach through half-begotten sod and primary scorch, wooden
sigils raised and reach for me when the sentences don't align,
your voice risen within wings flapping, wind and bells chime
behind the rain. I'll wait in the blue between flashes of lightning

inheritance folding all light away then toward my eye like
a Roman building in the middle of a rolling green meadow
mountains shatter to water inside me, I wait, like gravel
and granite, grey smoke and steam, the banks and the river fell,
I wait until I can close my eyes against each dawn, to stream
a valid defence at last when all communication retreats

; a spectrum where colour appears to wait for me

BALLANDEAN PYRAMID
Queensland, Australia

heavy water deep underground
    catches particles speeding to move
        solid mass, catching light, an initiative or

            a dream of lush grass, cold springs coiled
                during rigid gasps, easing the thaw;
                    our initiative to empty dark plots

                props to plot points on a graph,
                  dim lichen like light recoiling to envy;
            I should probably be better read

      this poem environments because
    new patterns grow sharp,
shrapnel fields sustain moss

slightly slid off-centre and to the left
    or south, tropical, greenery dreams
        vacating tautology, forests of youth

            journeying along an equator built of rafts
                – leave me, leaves are falling, I drift
                    against gains, grains settle upwind –

                speech-lead balloons drafted headward, winds
              give shape to verdant, floral architecture;
            I wait, neither lost nor found, for another

    – compared to a word for meadow, cognate
   of yellow-green, it seems speechifying
  in search of an empty, leaderless field,

  some scrub grass, some deranged wheat,
    bent stalks snapped in the wind, rain
      washing it all into whatever ditches,

        clichés march, hoisting flags, snapping
          flagged stones trodden underfoot
            weathered at the corners and down –

        drowned caulking washed out to sea,
        sewers with sawdust shovelled clean,
      I dream cobbled streets meeting my feet

    – what can last can last and what can't last
   can try to avoid being left behind our retreat
  but really how to judge each apostrophe,

  each apostatic bump along the line
    moving toward each new letter,
      hoping against hope to find the last

        grasp an embarrassment of riches,
          richness in the field, an agony against
            clasping newly mulched dirt to my chest,

        chastising each personal underworld
        anti-green circumflex, water unwhirling,
      domed ceiling upon which I now stand –

I withstand the full weight, waiting
until it is safe enough to move,
to mount a movement, arm in arm,

standards flapping, entrapped
platitudes ring hollow, capture
ears and hearts, meat caught

well-wrought and frigid prose,
butchered gothic bridges span
hesitant scratchings and matter

enspanned, conjoined, distraught
thought reaches, and in reaching
always finds that for which it solves –

loving great pastures, verdant chance
augured past surety or security against
sprained prosperity, strained future –

in my time we ran, what was it that ran,
I stretched my body across thought,
reaching as I was toward alien charms,

realms unstaffed by gregarious chants
implanted greed as soon as I grasped
desire undoing, and in undoing I find

a kind of pleasure undifferentiated from
feral, unbroken-because-never-whole
arch-humanity warming on the grass –

unfastening technology from an anchor
adrift within multifaceted surface flora,
grasping becomes naturalized, Kubla Khan

lives form nuclei, an atomic need
    meets satiating future imagining
        destiny in random images that

            speed by, unslowed, unfettered
                or unyoked, energy unsought
                    in my utterances I seek

                speaking plainly, playing newly
                greened dithyrambs, lyres
                nimbly leaping registers,

        wires snaking lime-green blades,
    emerald shadows cast as enviable
hues across my upturned face

displaced, deranged, unpasteurized
    lives gathering points toward climax –
        crescendo and denouement achieved,

            bleeding green, seeping each step,
                shambling toward ever-receding ecstasy,
                    forests, verdigris, skeletons to our great need –

CONOLY'S FOLLY (THE OBELISK)
County Kildare, Ireland

'His eyes will be yellow.'
yellow-ammeryellow-bellied
yellow-bellied sapsuckeryellow-bellied sapsuckersyellow-billed loon
yellow-birdyellow-breasted chatyellow-brick road
yellow-cardyellow-eyed grassyellow-eyed penguin
yellow-eyed penguinsyellow-goldsyellow-green
yellow-green algayellow-hairedyellow-hammer
yellow-legged tinamouyellow-legged tinamous yellow-necked mice
yellow-necked mouseyellow-rattleyellow-shafted flicker
yellow-throated yellow-throated warbleryellow 2G
yellow anemoneyellow bellyyellow bile
yellow birchyellow brick roadyellow cake
yellow cardyellow cardsyellow dog
yellow dog contractyellow feveryellow grease
yellow hordeyellow hordesyellow jack
yellow jacket
yellow jerseyyellow jessamine
yellow journalismyellow lightyellow locust
yellow locustsyellow menaceyellow onion
yellow orioleyellow pagesyellow perch
yellow perilyellow phosphorusyellow pine
yellow pocketyellow pocketsyellow poplar
yellow pressyellow pressesyellow rattle
yellow sheetyellow snowyellow spot
yellow spotsyellow terroryellow warbler
yellow wood anemoneyellow wood anemones yellow woodland
    anemone
yellow woodland anemonesyellowammeryellowammers
yellowbackyellowbacksyellowbellies

yellowbellyyellowbelly slideryellowberries
yellowberryyellowbillyellowbills
yellowbirdyellowbirdsyellowcake
yellowcakesyellowedyellower
yellowestyellowfinyellowfin tuna
yellowfinsyellowfishyellowfishes
yellowhammeryellowhammers yellowhorn
yellowieryellowiestyellowing
yellowingsyellowishyellowishly
yellowishnessyellowjacketyellowjackets
yellowlegsyellowlyyellowmargin triggerfish
yellownessyellownessesyellowroot
yellowsyellowseedyellowshanks
yellowshinsyellowtailyellowtail amberjack
yellowtail amberjacksyellowtailsyellowthroat
yellowthroatsyellowtopyellowtops
yellowwareyellowwoodyellowwoods
yellowwortyellowy

'Her eye wil be yellow.'
amber:
banana:
beige:
biscuit:
bisque:
blanched almond:
blonde:
brass:
bronze yellow:
butterscotch:
canary:
chamois:

champagne:

cornsilk:

daffodil:

dandelion:

ecru:

fawn:

flaxen:

gamboge:

gold:

goldenrod:

honey:

khaki:

lemon chiffon:

lemon:

linen:

moccasin:

mustard:

oatmeal:

old gold:

oyster:

platinum blond:

primrose:

saffron:

sand:

sandy:

school bus:

straw:

sulfur:

sunflower:

tow-coloured:

wheat:

wheaten:

'Here ye will be yellow.'

4:31
Coldplay – Yellow
Music video by Coldplay performing Yellow. © 2000 EMI Records
Ltd This label copy information is the subject of copyright protection.
OFFICIAL HD by emimusic | 2 years ago | 18,407,560 views

4:36
Coldplay – Yellow – Lyrics
Coldplay yellow and lyrics
by AliC09 | 3 years ago | 2,897,398 views

5:36
Coldplay – Yellow Live in Sydney 2003
Coldplay – Yellow Live in Sydney 2003 Early performance without
yellow balloon, but it's cant deny it as a great show.
by live4beck | 5 years ago | 4,944,691 views

4:05
Yellow by Neill Blomkamp
Neill Blomkamp was slated to become the director for the *Halo* movie
… However, plans changed and now he's the director for *District 9*.
by remisser | 5 years ago | 496,607 views

4:19
Coldplay – Yellow (Animation)
Suggested by user: bellaxmuerte5 Coldplay 'Yellow' Radio 1's Live
Lounge – Volume 2
by musicANDmuffins | 3 years ago | 2,174,521 views

3:51
Wiz Khalifa – Black And Yellow [Official Music Video]

© 2010 WMG 'Rolling Papers' @ wizkhalifa.com
HD by AtlanticVideos | 1 year ago | 117,924,981 views

4:30
Coldplay – Yellow (Live Tokyo 2009) (High Quality video) (HQ)
Coldplay – Yellow (Live Tokyo 2009) (High Quality video) (HQ)
Coldplay live Saitama Super Arena, Tokyo February 2009.
by TheLucho108 | 2 years ago | 1,744,422 views

4:59
Sara Bareilles – Yellow
Sara Bareilles – Yellow [Coldplay cover]
by itsalluncharted | 1 year ago | 276,749 views

2:19
Color YELLOW yellow song - Kindergarten
song by Frog Street Press, video by me
by beeblebrox01 | 2 years ago | 117,812 views

3:36
Mellow Yellow by Donovan
Mellow Yellow, by the lovely Donovan. You gotta love Donovan.
Dedicated to Vicki (vickim52), for requesting that I post this song.
by paulmccartneyrocks | 4 years ago | 869,832 views

4:49
Wiz Khalifa – Black And Yellow [G-Mix] ft. Snoop Dogg, Juicy J &
T-Pain
© 2011 WMG www.wizkhalifa.com
HD by AtlanticVideos | 1 year ago | 35,701,127 views

object
inhaled concept made
a remedy through science our agents
await Haunting a curved line or
costs imagined business as plans and marketing graphs Word meld
image and word scent overwhelm whatever container for my great need Orange

cinnamon
cloves peel citrus barrel and
box and basket oh my oh my explorer Kept
registered sentenced perfume arranged profits closed eye
to map our stores Calculated managerial tree shunts fresh water from a
refrigerated graph to say citrus I'm all anxious for you and depending so much on your

reply to
customer data Can I
get an advance copy of your book or
sneak peek or off to the races to riot again Sweat
seeps into pages a mortal quest to arrange a period with ghosts
periodicity Mobile durations and intimacy harvested against decay and squatting

ruins
　　Build out from the
　　　　cigar factory in a giant ring without
　　　　　　touching anyone anymore
Sweepers beat us to the punch before we had a chance
to spike it with rum Ruined as we were for our holiday when it rained the whole

time
　　in the amazon Deliveries
　　　　couldn't even get through the tangle of
　　　　　　snakes we left in our hotel lobby Maybe we need to sleep
in order to appreciate flower arrangements Made service cleaning
all that hydrangea dust into a performance piece y'know? I paused in the moment before diversifying my revenue
streams and remembered you

Stately
　　courts suffering duress
　　　　present what I do and who I am to the world
Images mashed into faces of loved ones wrecking homes for
each other A ravishing imagination revitalizes all of my experiences
and interests into politics Specialties aired and affirmed with community representatives

agents
online Await my calls
in living rooms beyond bewildered bed-
rooms dens and basements I call to them in my orange
voices with wild complaints channelled by the wind And invite their
advocacy on my behalf either toward or away from deep issues But not both never both

object
at the cursor that bounds
each further thought caught in barricades
Unable to shift or lock into some kind of capital recursive
rescue bought with credit I'm words apart from following or vindicating
whatever social purchase exists Between a high concept and lowly waves emanating from my secret pumpkin chest

Peels
and walls around
each stake provided by special interest
company directors like I get that they are into revolution
and are prepared to make my dreams a reality but what are my dreams
Vehicles purr in the dark husks from an imaginary war evidence that light carves lines

object

Cantilevered to process

these forms must be completed three times buddy

Oak tablets and chairs unlock to my face spaces I show by dashcam

in dying amber light Such unbridled hesitancy resonates as currencies rise and fall

The board

meets each autumn to determine

admittance at a secret location next door An easy

method for trimming digital wallets gilt free virtual placards announce me

What slow creeping dread arranges the pillows and makes the bed each morning

What catastrophe arrives with each breath gasping in the cold season

PETERSON'S FOLLY (SWAY TOWER)
Hampshire, England

                    It's a matter relative
Along the spring, autumn leaves fall then slowly wash with trash down-
stream of the signal tower. Where our hidden spot at the rapids discloses
frequency, the flow remains away from us, yet still bounded. As pertains
to flow, I try to stand and put my arms around you, but somehow the
moment passes without either of us knowing or moving. The end, at
last, wedges itself into our conversation and brings us closer.

                    To perspective, our living
                As we were before or
                After moments. Light
            Spreads matter across
            Our bodies, light waits.
        It's a matter relative
    To perspective, our living
We might ask when it began, or even what it was that began without
us, then again we already know, as part of the river knows, the banks
that have foreclosed any other purposeful direction except a trickle or
tributary, here or there. We save our leaves as they gather in small
pools downstream, wondering when and where they originated and
thankful that they exist. We gather in our arms to wait.

    As we were before or
After moments. Light
    Spreads matter across
        Our bodies, light waits.
        It's a matter relative
            To perspective, our living
            As we were before or
This hesitant method of scratching lines and dots, curves and crosses
into whatever surface, some dry mud, some bark, always forward;

regardless of the way we organize the marks, we read them in one path. The weather we owe each other, the winter for which we wait and organize remains all around; the most economic route into our lives avoiding brackish water, smoke and stagnant pools, increasing in size until all we can see is the natural economy around us and in all of our intimate communication.

<blockquote>
After moments. Light

Spreads matter across

Our bodies, light waits.

It's a matter relative

To perspective, our living

As we were before or

After moments. Light
</blockquote>

Cycling, break, and reform in the long moments before winter arrives, we turn and lean against our bricks, wood, and double-pane glass, our accumulated armour. That lasting second before frigid wind snaps branches, before sluggish ice-filled water passes, that last second that haunts every following narrative we paste together and anchor along the shore.

<blockquote>
Spreads matter across

Our bodies, light waits.

It's a matter relative

To perspective, our living

As we were before or

After moments. Light

Spreads matter across
</blockquote>

An afternoon in blades of light across our thighs. Blades of regular thickness slide along the ground to the river. Acting, actioning, navigating all meaningless vectors we see emanating from whatever original point, a wealth spread and moving across all matters we find important, our thighs, the river, our afternoon, to settle for even a moment on the banks. In the darkness and cold that follows, the banks cease to matter, their loam, scrub grass, and rocks fade with time, small shrubs fall and decay. We wait for all of this as each moment stretches into the future, each act of accumulation taking longer and longer to satisfy. We debate our past propriety and ease, our ownership carefully arranged with spaces between each word and sense marshalled by discrete marks. Trees fall, rocks avalanche onto our pathways, we didn't even know we'd begun climbing toward some resolute point against a windy crescendo rallying us onward and impeding any progress. We trail our dignity from the dirt to the peak, we fill our pockets with sandwiches, carrots, tomatoes, and our packs with water. We plan to wait out whatever winter approaches, whatever winter waits, our limbs heavy with the day's activities; our calendar turns into days and days, hours and minutes, and in a deep voice we proclaim our list of likes and dislikes into the descending twilight. We settle into our bones for another year on loan, holding.

> Our bodies, light waits
> It's a matter relative
> To perspective, our living
> As we were before or
> After moments. Light
> Spreads matter across
> Our bodies, light waits.

Passing each other and being passed over by the sun and the moon, we wait like wind. An unorganized moment into whatever present we find in the muddy banks and rushing river's seasonal throttle. Silted deposits accumulate and sludge their way toward collection, say in a book. We sit nearby, holding hands and thinking back to days when the water ran clear; an imaginary brook, false days when it was our water that ran clear, not just water, channelled by the banks toward whatever future we wanted, whatever lake or ocean we claimed. Reading into our relationship, as each moment passes, we stumble over words that were familiar and ask each other for clarification on simple matters of contract. Which twigs had we broken in our attempt to find the river banks? Under which branches had we cast our lot to watch the flow degrade from rapid to slow-seeping? Where can we place a door to shut against the sun's fake, measured plummet? What windows can we close to keep the wind out, and where will we find ourselves in the morning? Where are any of us against the light's last warm pulses growing slower and slower, receding into memory, leaving our faces flushed against cool, dark air? Where are we?

PART 2

IN RUINS WITHOUT THE APPEARANCE OF SOLITUDE

'See thou bring not to field or stone
The fancies found in books.'
— Ralph Waldo Emerson, *Waldeinsamkeit*

'I feel like a bridge about to collapse somewhere out in the back woods.'

— Bradford Cox

HENGE OR MESH: THE PRELUDE

I'm alone at dinner tonight,
single candle, low light,
elegant rain falls, is falling
above me, I am in the house,
hunched against best guesses,
figuring it all out or failing
still, like the rain falling
past my window, I dabble
in the known. I'm all
a-henge-or-mesh,
askew, heavy, slanted,
the country sides with me
and A sides with B before
C leaves the home to really
set my feet upon the path
like I own the alphabet with
energy dragged along behind
across illusory boundaries, a
photograph or distant scene
resolving from mist, I see or
saw grey slab to grey slab,
my path that I owned more
as leant rock monolith, not
a gathering of stones beneath me,
assembling at my whim;
a voice
alone
arrows
alike
arisen or

a raised
hands
a land
slides, a
down
driven or
driving
winds
swirl, a
circular
or encircled
stormcaller
enmeshment
broken henge
or breaking
or lately
a defiant or
wordfold
a bookfall
semiflip, a
gatherflash or
fallowfly and
graspafter or
walls, floors
and seams
gaps, gasps,
or gaping
wound or
unwound
sigiltrack
wordforms
letterspill

a positive
pause
without
caesura
a plosive
lightslash, a
gashflow
or spillforth
cloudburst over
millowfell to
firmfollow or
aughoanla
illusive dynamite or
illustrative damps,
a total system
over which
we all henge,
a narrow stem
draws water,
opens morning,
cobblestone road
lengthens my
journey, a load
a cheque or
an exchequer,
plunge or
plummet,
transfer or
transit tax,
all the way
from my fence
to the other fence,

faces along the hedge
like stone monoliths,
turning to change
every branch set to sway,
each twig about to snap,
every tree bent to read
faces atop the wall,
stone monoliths fall to pieces
every page thrown to forget,
each sentence about to fail,
every book trodden to dust,
faces line the horizon,
stone monoliths break into history,
every cable stretched to fray,
each screen about to flicker,
every website becomes stone,
other people raise their voices,
from afar, over the boundary,
calling the past a liar
as the sun crests noon,
a tidy pledge flung north
in between a moment
before lips meet flesh
and the chill breeze releases
an unkempt timeline
or status update faux pas
or fledgling text message,
a voice alone crumbles then
rises to say anything
begun today will fail,
there is nothing
and a house shape

spat into lyrics
vocal rain wall
a voice asking for bridges

lakeontario

cloudgown

mountainchin

pacificocean

silverbellow

elbowriver

rideaucanal

_ _ _VEIL_ _ _ _ _

sense, sure
| | t
        tense
a sentence or ten
nary | | ord | | a wall
symbols, mesh | veil a
sh | settles, ash // clouds

symbols, words terse
everywhere spare
not a thing w/ which
to think| instance
| | sur-e space a TMI *ancien* brain
rounded out by | brackets | breath
| earth curve shell ATM
warm tone | blown valeward |
disarmed, my body
ways and woods // legs
set afester| a baby cries
| trees rot, warm brown water

|which root
scratch | weeps
        |and exposed
scratch | tears
        |and exposed
coils me slitherward |
or exactly | as it is:was
a pile of |ants ..::!. mortal
|a thing | dissembling \

without interpretation|
| .   - _ . .   / /
/bird call and wild |
\cold, sterile |wind | /willed
|>geometry at the fragile and causal
limit | or < moth wing dust:|

                      |
Or > chorus raven comma | please
    plea flung patch < kora |
planetary swing -|- plea | >
    heaved skyward to   > |
say anything - against | >
    the said edifice - say > |
| < hitched forking| dir- >
    |ectory | branching two
    |by two for\ /radical
    |or rooted\/ patch
|furrowed |by sen \/se
    |that we\ /are all hiked|
|heaven |bound\/ and |
    |in it for\ /each other \

```
|sur | e  |
|  |pris|e  |
|pris|  e  |
|  |in  |
|| out  |
|  |er|  |
|er|  |
||ex   / it
|| ex |vail \
||   \    |
|  |  / |  |
|  |  \ |  |
|  | tion |
|ve| I |  nc
fn| l  |
| s | ce \
  n| d  ||
  d| vl |  |
||          ||
s|          | s
ing| igh | h
| glos  |s
|  |ing | s
||ed || s
ra|in   |s
in| sh    |s
| |ra|in    |s
| |ining  |h
```

listen:
lone cry
meadow
dusk

listen:
stone wall
barbed fence
dusk

listen:
bright star
patch of land
dusk

```
                        \
                        |
        |               |
      water -|          -|

                \
                    \
                        \
   damp forest floor        \
                        /
                      /
                    /
                  /
                /
                \ –
```

                              fields tilled
                          \ the hills | crawl -- |
         well | –   what hills | field
|comma| a start \ crawled, a rock wall crumbles into the stream
      |there| / | something falls | branches break
                              were | _I_ | water |
|valleys | | peaks – ponds |
        notches | t | roughs | aque-   \ land |
|grading | an escape \   ducts | gravity |
                              land| | fallow \ fields |
           |filling a | | season   break |
           |landscape | lesson | left

                                    53

                          tiny black | points | disappear
            footprints on | | paper
        green fields |letters | slaughtered
                            | | slanted light
right as evening | reeling | /
separates, us like   /
a joke from the   /
foreground   /
rocky slope   /
carves the   /
sky into   /
itsel   /
|its / elf
a / magnitude
/of ravens
 \
|or\ a closed
|circ\uit murmur|
   --ju\st like that
|the tree\s appear|
|a disposal|\
|of rights | \
|a forest | \
|in which | \
|no tree | \
|existed | \

secondary spread / __
flora and fauna |
technology -- |
and /wildfire |
where is the, the
volume button
/a rush upward
/a sunset bloom
/a landslide growl
/a tidal pool
/a tangle of moss
rather climb\
rocky steps\
rather claw\
back words\
| | rather
| | stop
| | than
| | distribute
| | rain
//without text

A PORTRAIT

gallery walls a-crumble
and I'm inside eating,

I'm inside eating my
lunch before work,

I'm eating lunch at
my work in the sun

and dust, so much
that a new statue

hardens around me,
impossible to resist

the accumulating import,
sunlight howls my thought
onwards from tight couplets
to a disregard for the dusty light
I follow and my body follows
and at some point I remember
the layers of shit on the walls
that seemed so important
while I was eating my lunch
in a broken gallery surrounded
by colleagues and acquaintances
before I had to start work
dismantling the portrait exhibit

TH THNG

smthng
strs thr, n
sttng thr
t sts

lrkng
bnd
t wts

ts hbt
t shft
shckld
t thngs n
rtcltd
prfrmnc

t f rch
stck
t bcm
rlzd r
t rlz
whchvr
hppns frst
wtrmrk
vr tslf
rms rsd

whspr
crss
fc

chst
hvs
wth brth
nd crvs
sch tmc
nmtn
brft wth-
t n nsd
th mmbrn
frght btwn
s nd t, wrds

sy thr t s
nd mn t s

rprs wth dwn
rnd th lttrs

dp lns crv n bd
th bjct, prfrmnc

gld br r
rnkld ncns
tch scrn r
tck bx

E I

oei
i ee, o
ii ee
i i

ui
ou
i ai

i ai
o i
ae
o i n
eiuae
eoae

ou o ea
o
o eoe
eaie o
o eaie
iee
ae i
aea
oe ie
a aie

ie
ao
ae

e
eae
i ea
a ae
u aoi
aiaio
ee i-
ou a iie
e eae
au eee
u a i, o

a ee i i
a ea i i

eie i a
aou e ee

ee ie ae a auae
e oe, eoae

o a o
ae iee
ou ee o
i o

THE THING REVEALED

SLOWING THE THING INTO FOCUS

words appear
shimmering in the air
somewhere above
and to the right of me
my forehead burns
the thing melts
into and of itself
a field undone, das Ding
salting the battlefield
between me and myself,
a punishment meted
for agrarian faults,
from A to C, primal
energy leaving home:
whether to be grown
or unfashionable, my son,
I hope you never know
the sun drags us all
through the galaxy
and we spin along
picking our pointless
battles, casting our
terrible nets wide,
hoping to find inside
some reason to continue
battling the void that forever
calls our name and gets
it all so entirely wrong

KING OF RUINS, RULER OF NOTHING

this is a myth about
the making of myths,
a myth about making
a myth, I tell myself:
it's a myth taken to view
itself as a monolith,
and I begin this poem, like
any other myth, with a character
sketch of a white man feeling inferior
and pretending himself a victor
of an imaginary war, asking himself:

what does politics
smell like first thing
in the morning, dawn
light tangled in her hair?

what keeps answering
the phone when plain
speaking gets called
into question again and
again? how can we say
what it is not and pretend
we can't say what it is?

start with the home: a
book calls to me from
my shelf, unopened.
okay, I say and power
up my computer to play

my new videogame that
cost the same as three
brand new poetry books.

a castle built upon
a rumbling subsurface
and corners start to pull
apart while the floor rolls,
walls crumble, what-
ever shakes will shake,
and a trenchant critic (me)
stalks amongst the
overgrown halls (my mind),
pounding the walls
until the tower (my ego)
resounds with his (my)
power and all
flee, at least
that's how he (I)
sees it, finally alone,
when the art he longs
to criticize ceases
to be made by him (us).

permit me this indulgence:
a tower now useful
as a lesson, an eyesore,
empty and purposeful as
a marker to avoid
the faults below,
and therein
words, when last

uttered, failing,
but in what way,
and failing whom?
words, when breath
and sound clip, haunting,
but in what way,
and haunting whom?
words, when most
effective and necessary,
taken in what way
and taken from whom?

the passive voice shambles
over and flops on the rug
below an old, dusty hearth
with pictures decorating
the mantle, primitive images
in their desire for realism.

block and bit, byte and wave
gathered from the material
wrecked all around, a past.

a sentence gathered your hair,
your shoulder, your neck
and chest, your arm and
face, your eyes, ears,
mouth, and nose, your
forearm, your calf, your heart,
your thigh, your groin assembled,
and now what monster have we?

to be unradical in the face
of boredom, to resist novelty
for the sake of advancement,
to turn and turn again
from intellectual fads,
when asked to move forward,
neither sitting nor standing,
instead dissembling as questions
surround, or curiosity
manifested as demand
comes calling, to remain
resistant to the urge
to make it new, to find
integrity in that resistance
and to remain unwilling
for your pleasure.

no manifesto marketing,
don't say no, don't say
anything to anyone,
kill your soundbite,
just do it, drink schweppes,
don't climb the awful
heights of your
sovereignty
to look down.

note to self, dear friend:
collect some driftwood,
pick up some concrete,
build something nobody
needs or wants and

get on with living
your life, try not to
destroy your others
into crumbling bits
with the sheer weight
of dealing with you.
in the meantime
accept the irony
and hypocrisy
of yourself and
embrace the imperative.
thoughts are what
your mind shits out
and as you get older
you'll realize just
how important
it is to be
regular.
sincerely yours,
Jason

at the hedge lining my property,
the hedge that used to line my property,
at the stucco around my house,
the stucco that used to hold the world out,
at the switch on the television
or radio that used to stop the signal,
at my eyelids which close
to recoup something from the day,
there must come a point
at which to say no further,
if only so that I can sleep

without purchasing that time
from myself free of commercials
with my bonus minutes donated
by my employer/family/govt.

a series of platforms shifting
in and out of the poem, into
a series of platforms crawling
into a space below all other
buildings, unstructured platforms,
a counterpoint to all of my
anxious desires for control,
an unarchitecture for thinking.

a lot of poets talk about
involving the reader
in making the poem,
in making meaning,
and I guess that's
a nod to some theory,
but I've always thought:
wouldn't it be great
if I could get the reader
to help me clean my garage?

it takes a village aware of itself
to make someone famous
and it takes another's village
to make someone rich,
but it takes a village in the first place
to really hone in on your brand.

The Charm

'In any case, whenever there was a chance to publish a small pamphlet or book, my temptation was to cut from it any poem that did not seem to me then and there to make adamant sense as a poem, and consequently I tended to ignore a kind of statement in poetry that accumulates its occasion as much by means of its awkwardnesses as by its overt successes.'

                                          – Robert Creeley, preface to *The Charm*

'I believe that since the Industrial Revolution western questions of value are sardonic, if not sarcastic and that my only resource as a poet in 1982 is to put myself on the side of things which exists at an angle slightly askew to any desire for fame, or even value for the "works," forget about value as it's perceived, and take as much pleasure in my life as a poet as desire can construe and hurry to change the world in small performance as others like John Cage have done, since you can't stop fucking writing anyway.'

                                        – Bernadette Mayer, Mimeo *Argument*

REMEMBERING MOUNTAINS
for Jordan

a lip in the stride or waterproof backpack in the warm sun to quest
seals over sand grain by grain toward the tide slowly sweeping the
beach to sea, the beach to sea, an ocean undone by tectonic plates,
plates shift and flip into the world, making the world a staccato series
of ragged blips playing into space, the atmosphere lowering and
falling for months, a mouth of cloud cover occluding time, like chil-
dren playing in the creek, like the eventual summer sun, a boulder
territory, a territory made audible by gull beaks clipping rock into
dust, dust formed by cracking mollusks, by leaping from ground to
sky, a sky creating a register to populate with more sound, more
sound and then the world rushes in with laughter

EJECTA FLORA FAUNA
for a.raw

simple projectile garden
open-plan walkway with
water feature, sculpted
elegance pretends nature agrees

print certifies a present fixity
enough sound to remove staying
escarpment folds land
sky land bowing again

slug forward, bright noise, shift away
each wave with noise atop
breaks against each day
we struggle through until dawn

a bell ejects sound in a ring
around the centre, a calm edge
circular wave ebbing and cresting
dissipating outward, carrying relief

without a land to land upon
vowels shift and quake becomes
quiver or wander glass over time
to a transparent necessity we share

insert molten or lava or rock flow
where a shift from beware
becomes curiosity and home
balances against nostalgia

adrift, maybe, and yet anchored to that
eternal slip and tide, pull and release,
a great magnetic ash cloud billows
neither formless nor absent, unwhole

waves propagate into crescendo, but who knows
what plucked string began such
cacophony and which unantagonistic
beach loses no sand to history

I'M PUTTING ALL THE PUNCTUATION
BACK INTO THIS POEM
for Sachi

maybe not in other poems, but
I find that in this one the lines
muddy and blur too much. usually,
I like that effect and manipulate it
to subvert and highlight whatever
it is I think I'm talking about, but,
since it is just us two here,
I feel like I need to be a little more
direct and perhaps kind, as Sachi
once put it. I had been looking
for a way to infuse my poems
with kindness, to approach each
poem with the goal of layering
kindness underneath like bedrock
or underfloor heating in a cold
part of the world, but sometimes
my cynicism and training kick in
to defuse that urge and redirect
my efforts toward theory, or
a mistrust of the olive branch, so
I'm putting all the punctuation
back into this poem as a gesture.
we'll see if the desired effect
can survive my tinkerer's need
to whittle and pin, to hammer
and weld, to gather bits and
bolt them together, to repair, or
to look at everything like it
could use a bit of tinkering.

AS THE SCULPTURE TIPS
for Jake

left un-
resem-
bling glue
and dust, sun-
light cuts as if
it could cut itself
a shape against
the background, in
relief
I see words
as the sculpture
tips, then shatters
on the marble
tiles, in the white
gallery, near
the city centre, and of
the spectacle we
will say:
which
hoser let the guy-
wire slacken, and
which poet wrote
'a relic, a relic
my horse for a relic'?

YOU ARE NOT A CRUSTY COMET
for me

When I saw an article about an
object that was visiting our
solar system from somewhere
beyond, I immediately thought
of the title for this poem, and
then when I started thinking
I should write a poem with
that title, it suddenly occurred
to me that I should dedicate
the poem to myself because,
while I am not, in fact, a crusty
comet, no matter how much
I identify with our interstellar
visitor, at times I have felt
like a cold, heavy, unknown
rock tumbling willy-nilly through
the vast vacuum of space, but
in reality I have a great job, two
wonderful children, and an
incredible wife. I like what I
write, for the most part, and
I am lucky to have a few friends.

FLANGE IS GOING TO DO YOU GOOD
for rob

metal bit to metal bit,
I wanted to say that
there are too many
of us in this trash heap,
today of all days.

wooden thing that holds
two other things together,
the unmentionable joint,
the void that binds,
light connecting to solid
objects, getting that space
between to delineate a curve
toward or away, joined
by perspective, a unitary
point somewhere between.

there is the beginning and
the end, when all is said
and when nothing is done.

what I wanted to say to you
was that I finally got to use
the word *flange* in a poem.

WE'RE NO LONGER SPEAKING
for nobody

we are no longer speaking.
terms on which we might
continue speaking include:
a bundle of wool sweaters,
trees recently felled –
get specific, they say,
and who, I say, is they,
they say I say, and so
we are no longer speaking
in terms that reflect
our disposable coffee
cups, plastic bottles, and
pink hair curlers, our dirty
diapers, curdled milk, and
mountains of cigarette
butts, our empty graves
with all that mud piled
beside, and our rotten
carrot mound, packages
and packages of blanched
almonds we never knew
what to do with, and
our GPS device, old
flip phone, VCR, and
cassette tapes, fruit
flies, Pac-Man T-shirts
we barely wore, down
we stumble and we realize
we are no longer

speaking to each other,
only to the virtual assistant
hanging on our every pause.

ADAGE AGAINST SELF-CONTROL
for Stephen

In this time-honoured manner of taking care
not to offend, or of voting for irresponsible
parties, I do solemnly swear to adhere to
the adage against self-control, that it might
sway me to purpose against all my self-
doubt and debt, my unworthy sunrise view;
a ragged right of the marginal to declaim
the loudest into whatever face assumes
control, and in whatever media is to hand,
voice, sight, sound, or smell;
                              a basic reward
for lining the birdcages where the gentry,
in fledgling form, gain feathers, moult, and
feebly take flight for the first time, a pittance
really, because when the cage is removed,
the birds fly away while the rest of us squat.

THIS POEM WAS ONCE IN A SECTION CALLED CUSTOMER SERVICE
for Ian

then I changed the sections and the gist of the book
and the tone of a section called *customer service*
just didn't fit, still I wanted to keep something of the
sentiment of the poems I cut so here goes: money and
status are corrupting and bending all of our best and
brightest, social media is creating a generation of shoot-
from-the-hip thinkers quick to vomit opinions as fact
and regurgitate headlines and click-bait, and zero
opportunities for long-term, stable work are degrading
our middle-aged and elderly citizens until they lose
all hope and succumb to mind-numbing labour instead
of retiring with respect and dignity because they
fucked up and counted on the examples from their
parents' generation and missed the reality around them

BRIGHTEST GALAXY IN THE UNIVERSE FOUND
for Natalee

it's Tuesday morning
and I'm drinking coffee –
it's good coffee for a
Tuesday morning

I'm listening to a new
album, specifically
a song about guns
and drugs and ego and

my kid has a fever
but, hey, astronomers
have found the brightest
galaxy in the universe
so I guess I'll have more
coffee and watch a cartoon.

I'm having coffee on a Tuesday
morning and watching a cartoon
in which a magical dog and
a young boy learn respect
for themselves and it occurs
to me that I am all my Others

my kid has a fever
and in some other galaxy
some other kid probably
has a fever and might
just be discovering
respect for his Others.

THE DAY THE UFO STOOD STILL
for Natalie

a day like any other,
really, rain dropping
centuries of data
to break against
my windows, water
streaming down glass
like music over wifi,
and I don't understand
much of what *world*
means anymore, or
what it ever meant.

millionaires made
off whimsical tithes:
click, here's 99 cents
for your dancing
and singing, it builds
to a crescendo and
sustains around noon
with eggs frying
in butter on glass,
in a non-stick pan,
in a new condo
building designed
to look like an old
condo building, yet
the letters don't
settle, they shift
and seethe just
under the surface –

a layer below
cloud cover,
a lining laying
a better day
at our feet.

FILE MANAGEMENT
for Nicole

skip ahead ten seconds
there is the sun again
now back about three days
it rained from low clouds
pause the record mid-song
what the magnetic tape
wanted to tell us was to
skip ahead ten seconds
a pulse and screech
the floppy disc jammed
financial data in ruins
an archive inflamed and
stuck in time forever
pause the movie file
there is the sun again
let's not go outside
instead we find photos
stored in the cloud
of our parents' wires
from our childhood
of our children
from our past travels
of our places
from the things
of our life together
skip ahead ten seconds
that is exactly it there
the moment when
we fell in love with
our capacity to store

I AM NOT A YOUNG MAN
for Andrea

it isn't Friday night
and yet I'm having pizza,
my love, without you
in a strange town and
although it isn't Friday
night, I hope you
are having pizza too

the dim lights, horns
muffled, the lights
are dim, aren't they?
or my eyes have gone
and my back aches
and I can't hear horns
brassing in the song
the way I used to

I am not a young man,
my love, except
when I'm with you
listening to music
and eating pizza
on a Friday night

a litany for [cursed object]s

'The Malice of Inanimate Objects is a subject upon which an old friend of mine was fond of dilating, and not without justification. In the lives of all of us, short or long, there have been days, dreadful days, on which we have had to acknowledge with gloomy resignation that our world has turned against us. I do not mean the human world of our relations and friends: to enlarge on that is the province of nearly every modern novelist. In their books it is called "Life" and an odd enough hash it is as they portray it. No, it is the world of things that do not speak or work or hold congresses and conferences. It includes such beings as the collar stud, the inkstand, the fire, the razor, and, as age increases, the extra step on the staircase which leads you either to expect or not to expect it.'

– M. R. James, *The Malice of Inanimate Objects*

'We cannot stand in full daylight
and see the objects in the dark.'

– Lucretius, Book IV of *De Rerum Natura*

'As a general rule I try not to get overly emotionally invested in objects, and there are certain types of them that I particularly try not to get attached to. It's better not to think of "owning" sunglasses, headphones, or bikes, but to consider them on loan to you until they get broken or stolen.'

– from somewhere on the Internet.

[chiasmus, parallax, ships passing in the night, reading the book of
spells on my phone at night in bed when it should be dark, stars
falling outside and literary devices drawing all manner of beasts,
waiting, preying.]

my phone calls to me,
I press the button and
hear an objectless voice:

Jewels, ornate stones, and
broken bones, names will
never suit me, or so the
saying goes: player's choice.
The weather today will be
cloudy and rainy. As I listen,
I close the window against
thunderclouds, assuming
my sovereignty to be intact,
then I make a bid for the crown
during this game of Temple Run.

I falter even as my new [cursed object]
tries to describe me as tithed to the sea, or
some bizarre password known to the wee
demon hidden in my thin metal case behind
a glass screen laughing at me trying to
activate my home so my assistant
can connect the thermostat with
all my speakers tucked into the ceiling,
with the locks on my door, and
security cameras, with my oven,
toaster, and refrigerator, my router gasps;

I'm myself for my own benefit and
carefully describe my being as
a fingerprint, a voice print, a
heartbeat, some breath, a
phrase I use most often, a job
I had, my mother's maiden name,
some number I got for being born,
I mean, I'm barely anything other
than the data I've accumulated
in the eyes of a state or company
desperate to know me.

Ok, [cursed object], search: *The Prelude.*

Jason, I found the following:
I led my horse, and, stumbling on, at length
Came to a bottom, where in former times
A murderer had been hung in iron chains.
The gibbet-mast had mouldered down, the bones
And iron case were gone; but on the turf,
Hard by, soon after that fell deed was wrought,
Some unknown hand had carved the murderer's name.
The monumental letters were inscribed
In times long past; but still, from year to year,
By superstition of the neighbourhood,
The grass is cleared away, and to this hour
The characters are fresh and visible.

Consider attending a screening of
Beauty and The Beast *starring*
Emma Watson in a story of love
overcoming the uncanny valley.

Are you still there?
Tap to download
the_prelude.pdf.

Say aloud: what is
the weather today
in Phoenix? *Did*
you mean to order
The Weather *by*
Lisa Robertson on
Amazon? Yes. Say:
I need groceries.
Say: hold me, I've
fallen and I can't
get up. Say: return
to me my lost love
and undo all the ills
that have befallen me.

From my bedside table,
my [cursed object] speaks:
Look at my screen to unlock
secrets within that
connect you to the
universe, stare through
me at the stars and dream
while I count your heart-
beats to keep you alive
and to keep you within
my system, bound as we
are to each other now
by the contract you
signed and for which

you pay your price, while
serving you, at your behest, I
will burn your willpower,
watch me shine even as your kids
topple off the coffee table.

Subscribe to the West Wind
ad-free for ten years and get
a keychain with a picture of
whatever you hold most dear
on one side and one of six
limited edition West Wind
logos on the reverse (you choose).

Hi Jason, I see you forgot
your password again.
Your password prompt is:
what was your first job?
While you are remembering
what it was, you might
also consider checking out
these delightful everyday
objects recast in expensive
materials sold exclusively
at Tiffany's online.

One-thousand-dollar
tin cup, what are you
to me, who has
fallen so low as to pay
all my attentiveness
one app rating at a time.

Oh, one-thousand-dollar
tin cup from Tiffany's,
what mortal censure
or heinous gaffe invited
you into existence?

Sure, I'd happily watch
a video to earn enough
credits to keep typing
this poem on my on-
line word processing
program.

I look to you, oh one-
thousand-dollar tin
cup as if you are
some vulgar heuristic
with which to divine
true happiness as a
world declines into
being after years of
becoming. Dear [cursed
object], guide me home,
I've fallen off my horse
and I'm confused.

*Press the button
and pay $1.99 to
hear more about
the adventures of
the [cursed object].*

Press the home button
to return to the main
menu.

Perhaps the most
famous of all [cursed
object], hope
dies last, yet torture
and despair haunt us all
in its wake as it ruins
whatever story
it touches through
history and fiction
as one guillotine blade
after another severs
our sovereignty, we
acknowledge that our
usage of this [cursed
object] includes
in-app purchases
by clicking the
ToS checkbox.

Say: the saying goes
'suits me,' or so the
names will never
be broken, bones,
jewels and stones –
this is a romantic
song to sing alone
at midnight along
the road, road-
side assistance twenty-four

hours a day at the
push of a button,
just say: help me,
I'm loneliest while living
life instead. Say: I lived
entirely within the said

Whoever shall open
the seven-sealed box
surrounded though
the jewel be by signs
and charms, heed
and throw it to sea!

By continuing to read or listen to this poem
you acknowledge that we use cookies for reasons.
Click *learn more* to be further confused by legal
nonsense coupled with technical jargon, or pay
the low, low price of $9.99 a month to continue
hearing or reading this poem without your
eye colour and size, weight, height, shoe
size, friends of friends' interests, and
vacation photos being recorded, analyzed by
our team of underpaid nerds, and ultimately
shared with our marketing partners to extract
every ounce of profitability from your nostalgia
expressed by this interaction and to learn
how to further erode whatever willpower you
have left over from the numerous poetry readings
or books you have encountered in the past
with similar warnings you failed to heed.

Squeeze the sides
to activate your assistant,
stroke the right headphone,
or say: ok, [cursed object],
say yes, say
search for
a voice
at night
alone, say
arrows.

saying aloud:
[cursed object],
play some
relaxing music or
tell me a joke,
read me a story,
entertain my children,
cast the new episode of
The Good Place to the
TV upstairs, no *Golden Girls*,
record my favourite moments
and organize them for me,
take my pictures,
take my videos.

I've got so many clips
of the same guy trying
the doors of our cars
in the middle of the
night that I should
make a scrapbook.

[cursed object], could
you do that for me?

Please note, reading this
gives the app access to
portions of your memory
to provide you with a
fully customized social
experience for the low,
low price of $1.99 and
an additional ten bucks
to remove advertisements.
We hope your VR experience
was better than real life!
To uninstall, please call
customer support between
9 and 5 Nepal Standard Time.

Every year we say to
ourselves and each other:
the device is a barrier
that we will overcome
the same way we
pushed through country-
sides, over farmland,
through the walls of
our homes and workplaces.
We accept all forms of payment
including time, attention, willpower,
and in return we cleave ourselves
from whatever once connected us
to a sense of place and possibility,

and we tell ourselves, every night,
that other people are doing this to us.

As the blue glow from my screen fades
and I direct deposit all my paycheques,
I tell myself that I resisted something
because I didn't pay any money
to the people that made that game
with the candy, with the bubbles, with
the shapes, with the colours, with
the flashing lights, that I played
when I got married, when I saw
the Tragically Hip's last show, when
my sons were born, when I graduated
from college, when I had invasive surgery,
at dawn, at night, at lunch, on the beach,
in the bathroom, in my casket at my funeral.

Calling the past a liar
as the sun crests noon,
a tidy pledge flung north
before lips meet flesh.
Voices echo from empty
rooms, someone whistles
out of tune and out of sight,
perhaps upstairs when
nobody else is home,
some thumps and bangs
and cupboards open and
close and it appears as though
my smart home has lost it.

Jason, my [cursed object]
calls to me, *write this down:*
an unkempt timeline
or status update faux pas
or fledgling text message
hurled into the ether
trying to reach me
or my digital assistant
leave us a message.

I say a voice alone in the
living room rises
from the black cone
to say anything
begun today will fail
there is nothing
and a house shape
spat into lyrics
vocal rain.

Ringing or unrung,
a bell announces
unpunctuated wilderness.

The voice becomes
rhythmic, unbinding
tone from source,
unhinged spillage
echoing without
what sounded deep
from within, an
empty reminder

to buy more fruit
pings for nobody.

Paging or unpaged,
a book announces
interiority, lionized.

Covers flung askance
through a dark
doorway edged
by faint, thin light –

a solipsistic frame around
mind or beams and lintel.
As if waiting, accrued pages
gather dust over time atop
forgotten window sills
with each mote latching
one second to the next.

Waxing or waning,
a candle announces
darkness and proximity –

closed book, silent bell,
unlit candle and doorway,
silent page, muted metal,
wax with no wick, and
against this mirror
waste accumulates
under the nails,
between the teeth,

waiting until the flash
of light ignites each fleshy
page, melts wax, sets
emptiness to ringing
and the [cursed object]
looks back at me looking.

Jason, it says:
have you
heard
the one
about a doll
animated
some say
by a demon,
others say she
is the spirit of
a seven-year-old girl?

Anabelle
could move and
write help us
despite there
being no parchment
and no one home
and despite
being a doll.

Or what about this one?
Guilty of murdering
his father-in-law
for sitting in his
favourite chair,

Thomas Busby,
coiner and drunkard,
condemned to death
by hanging until dead
then suspended
in chains on a gibbet
at the crossroads
where the Busby
Stoop Inn stood.

On the way
to his death
he stopped at
the inn to curse
the chair such that
anyone brave enough
to take a seat
would be haunted
and soon die:

'May sudden death come
to anyone,' he said, 'who
dare sitteth in my chair.'

Everyone who
so foolishly sat
on Busby's chair
experienced severe
itching, paranoia,
voices from beyond,
confusion, items
moved in their
presence, written

warnings on walls
and mirrors, and
shortly thereafter
guess what happened, Jason?
Can you believe it? They died.

Hey, let me try something,
my [cursed object] announces:
between dim grey foliage,
punctuated dark canopy,
finger and hand skyward
closing the day, placing
night over our eyes
            us wanderers
defying angles for growth
warm wind and stars, trees.

The [cursed object] mutters,
passing me to the grave:
press the button to hear
a bird call
            a bird calls

[Praying, weighty, beasts in manner and kind on devices literally outside
fallow stores, dark as it should be when, recharging in my cradle, I lay
my poet down to sleep to awaken like ships passing in the night, parallax,
chiasmus.]

The poems in *Cursed Objects* orbit the idea of objects as distinct from me, not given their characteristics or existence solely by my will, thought, or existence. Technological innovations are challenging the status of objects as extensions of my sovereignty and the arrow that pointed from me at the centre to them all around me is now pointing both ways. I am given an existence by objects, literally courtesy of medical technology, or protection from climates that would be harmful, and more subtly in the sense that in some cases I am an accumulation of data to them. I am their human.

LÈSE-MAJESTÉ

The folly is an aesthetic embellishment of a structure, an object that is usually devoid of an architectural purpose beyond delight. Following in the ludic footsteps of the flâneur, except taking a more peripatetic approach, I use the nature of the folly to understand the artificial construction of a community which is also a kind of object against which an individual finds itself defined.

'Broadway Tower': I had in mind a drive in the late hours of the night when writing this poem. It has a very violet hue in my mind. It began with some musings on community. What gathers us around a flame in the dark, what connects us across distances and time, and how do we traverse the void we must cross to get to each other?

'The Casino at Marino': Here, the sound of the bells bringing people to town from the fields struck me deeply, and I wanted to replicate that in some way in the poem. Technology functions like a herald in a similar way, bringing us all in to a virtual space together, putting ourselves out there online, for example.

'Ballandean Pyramid': Structurally, I wanted the poem to mimic the shape of pyramid, steps up to a point, but I fell in love with the flowing shape it was taking and it became a poem about a journey. I imagined a raft adrift on a slowly moving river with somebody poling it along, pushing off the banks, heading somewhere but not an exact destination.

'Conoly's Folly (The Obelisk)': I copied the search results for the colour yellow and loved how they ended up all jammed together. The folly is an aesthetic embellishment of a structure, usually devoid of an architectural purpose beyond delight, and in some ways I felt search results without their context present the same kind of pleasure. The poem loosely follows the structure of Conoly's Folly with three levels, one of which is an obelisk. The building has stone pineapples on it too, so...

'Rushton Triangular Lodge': I seized upon the triangle as a structural unit in the title, but the mystery implied by the word 'lodge' snuck in too. With the words and lines breaking in unexpected places, no punctuation, and melding images together, I wanted to hint at the exhilaration of communing with the unknown. The triangular structure also gives the poem a propulsive force starting at the peak and exploding outward in a shockwave to the last line of the stanza.

'Peterson's Folly (Sway Tower)': Picking up on the word 'sway' in the title, I knew I wanted to capture that movement in some way. The building has an external staircase. The original builder found encouragement from the great beyond when he consulted a medium and contacted the spirit of Sir Christopher Wren prior to building.

'_ _ _Veil_ _ _ _ _': Punctuation acts as a kind of monitor for words, without it we'd have just a bunch of letters run together, maybe with spaces to delineate them. I wanted the punctuation to be more intrusive than helpful, sometimes completely overwhelming words and sense. I mostly used non-standard punctuation marks to highlight this effect and to draw in the connotations of their other uses.

'Th Thng,' 'E I,' 'The Thing Revealed': I wanted to place into ruins the idea that a folly derives its identity by not being functional as a building, that it is not a proper building. There is a third term that gives shape to the folly's duality. What is a thing without its accompanying words? What are words?

'King of Ruines, Ruler of Nothing': The arrogance and privilege involved in making something useless, like a folly, far outstrips the aesthetic pleasure derived either in the creation or consumption of the useless thing. Power for the sake of power is one such folly.

THE CHARM

The poems I wrote to several of my close friends and family in this section heighten the stakes of treating people as objects constructed of words. By reducing the people I love to the status of a reference, I create a bifurcated space where they become useful to me and where I can indicate to them their importance in my life. These might be the cruellest poems I have ever written because of that duality.

A LITANY FOR [CURSED OBJECT]S

In this long poem, I wanted to drag the object as other across the line of obscurity and dramatize the relationship I've developed with the devices in my life. In the sense of a repetitious series of petitions, I called it a litany because every day I launch my pleas at my devices, begging them to do my bidding. More and more, the devices are coming to understand me and my needs, anticipating what I will ask, learning how to respond to my requests (and in the darkest of moments, guiding my thoughts, bending my will, and directing me to things I don't want or need at a dire cost). By the end of this poem, I wasn't sure if I was writing it or it was being written for me.

The quote that I have included in this poem is from William Wordsworth's *The Prelude* XII (1805).

ACKNOWLEDGEMENTS

I wrote this book in many places, over many years, which means I have too many people to thank properly. Thanks to rob mclennan for always prompting me to challenge my tendency toward reclusiveness that manifests physically but also with writing. I'd be nowhere without my friends, colleagues, and writers I don't know personally, all of whom I turn to for guidance or support, whether they know it or not, since sometimes it only happens in my mind. In no particular order, thank you to: Sachiko Murakami, Kevin Marechal De Carteret, a.rawlings, Jake Kennedy, Amanda Earl, Jessica Smith, Natalie Simpson, Chris Ewart, Natalee Caple, Jeremy Leipert, Stephen Brockwell, Steve Collis, Lisa Robertson, Fred Wah, Nicole Markotić, and Louis Cabri. Jordan Scott read many versions of the poems in this book in many cities, and without his thoughtful questions it would have ended up a very different book.

Pieces of this book were published or performed over the years in many places as well and I'll do my best to include them all, but I'm sure I'll miss one or two and for that I apologize. Thanks to N/A, *above/ground*, Chaudiere, Zouch, Heavy Industries, *Poetry is Dead*, Cordite, the Rusty Toque, Peter F. Yacht Club, the Calgary Renaissance, and *Touch the Donkey*. Thanks also to the people who organized the Word Ruckus, Flywheel, Factory, and KSW events where many of these poems first encountered an audience.

For guidance, sharp edits, thoughtful discussion, and the example of her own excellent poetry, I owe Susan Holbrook a mountain of gratitude. Her patience, playfulness, and steady hand are evident on every page of this manuscript. I feel very lucky to have had the chance to work with her.

To Alana Wilcox, for fighting for us, and keeping the dream alive, I'd like to say a heartfelt thank you. I'm grateful to everyone at Coach House, each of whom plays a vital role in bringing dreams to life. You are all champions!

I thank my lucky stars that Sandy Lam encouraged me to meet her friend for a blind date many years ago. My writing, my life, my everything, every day we share, Andrea Ryer, I am grateful and thankful most of all to you. For your humour, intelligence, care, and courage, thank you. I wouldn't be the writer I am today without you and the gift of our family.